From Quasars to Gamma-Ray Bursts: Unraveling the Mysteries of High-Energy Astrophysics

Bergman

TABLE OF CONTENTS

Chapter 7: Supernovae and Neutron Stars   51

Chapter 8: Black Holes: The Ultimate Powerhouses   59

Chapter 9: High-Energy Astrophysics and Cosmology   67

Chapter 1: Introduction to High-Energy Astrophysics

The Fascinating World of High-Energy Astrophysics

Welcome to the enthralling realm of high-energy astrophysics, where the cosmos takes on a whole new dimension of awe-inspiring wonders. In this subchapter, we will embark on an exploration of the mysteries that lie beyond our earthly boundaries, delving into the captivating domain of high-energy astrophysics.

Astrophysics as a field of study involves the application of various scientific disciplines to understand the vastness of the universe. However, high-energy astrophysics specifically focuses on the extreme phenomena occurring in the cosmos, where unimaginable energy levels and mind-boggling celestial events take place.

One of the most fascinating aspects of high-energy astrophysics is the study of quasars. These incredibly bright and distant objects emit immense amounts of energy, making them some of the most luminous entities in the universe. By unraveling the secrets of quasars, scientists gain valuable insights into the early stages of galaxy formation and the behavior of supermassive black holes.

Another mind-blowing phenomenon within the realm of high-energy astrophysics is gamma-ray bursts (GRBs). These intense bursts of gamma radiation, lasting mere seconds to minutes, release more energy than our sun will emit in its entire lifetime. GRBs remain a puzzle to astronomers, with their origins and mechanisms of production still not fully understood. Studying these cosmic fireworks can shed light on the most violent events in the universe, such as the collapse of massive stars or the collision of neutron stars.

In our quest to comprehend high-energy astrophysics, we must also explore the realm of cosmic rays. These highly energetic particles, originating from sources both inside and outside our galaxy, bombard the Earth from all directions. By studying cosmic rays, scientists hope to unravel the mysteries of their sources and the mechanisms behind their acceleration, which could provide invaluable insights into the workings of the universe.

High-energy astrophysics is a captivating field that pushes the boundaries of our knowledge and challenges our understanding of the universe. Its discoveries not only captivate scientists but also have the potential to impact our everyday lives, from advancing technology to improving our understanding of the fundamental forces that shape our existence.

So, join us on this exhilarating journey through the fascinating world of high-energy astrophysics. Whether you are an amateur stargazer, a seasoned astrophysicist, or simply someone with a curious mind, this subchapter will ignite your imagination and leave you in awe of the wonders that lie beyond our planet.

Historical Overview of High-Energy Astrophysics

Astrophysics has always been a captivating field, delving into the mysteries of the universe and providing glimpses into the workings of celestial objects. However, it is in the realm of high-energy astrophysics that the universe truly reveals its most extreme phenomena. From quasars to gamma-ray bursts, this subchapter explores the historical milestones that have unraveled the mysteries of high-energy astrophysics.

In the early 20th century, the study of astrophysics was primarily focused on the visible light emitted by stars and galaxies. It wasn't until 1932 that the first cosmic rays, highly energetic particles from space, were discovered by physicist Victor Hess. This groundbreaking observation hinted at the existence of high-energy phenomena beyond the reach of traditional telescopes.

The discovery of the first X-ray source outside our solar system in 1962 marked a turning point in high-energy astrophysics. Astronomers soon realized that X-rays and gamma-rays, with their much higher energies than visible light, could provide vital information about the most energetic processes in the universe. Exploratory missions, such as the Uhuru satellite launched in 1970, opened a new window for studying high-energy astrophysics.

The birth of X-ray astronomy gave rise to incredible findings. In 1967, the first X-ray pulsar, a rapidly rotating neutron star, was discovered. This revelation paved the way for understanding the behavior of matter under extreme gravitational forces. Subsequently, in 1974, the discovery of the first X-ray binary system, where a compact object accretes matter from a companion star, provided further insights into the dynamics of high-energy processes.

The 1990s marked a new era with the launch of the Compton Gamma Ray Observatory. This satellite enabled astronomers to observe gamma-ray bursts, the most energetic events in the universe, for the first time. These fleeting bursts of gamma-rays, lasting only a few seconds, puzzled scientists and sparked a race to understand their origins. It wasn't until 1997 that the first optical afterglow of a gamma-ray burst was detected, leading to the realization that these cataclysmic events were associated with the explosive deaths of massive stars.

With the advent of space-based observatories like the Chandra X-ray Observatory and the Fermi Gamma-ray Space Telescope, high-energy astrophysics has continued to push the boundaries of our knowledge. These observatories have revealed the existence of supermassive black holes at the centers of galaxies, shedding light on their role in the evolution of cosmic structures.

In conclusion, the historical overview of high-energy astrophysics demonstrates how humanity's quest to understand the universe's most extreme phenomena has unfolded over the years. From the discovery of cosmic rays to the detection of gamma-ray bursts, each milestone has brought us closer to unraveling the mysteries of high-energy astrophysics. As our technology advances, we can expect even more profound revelations that will reshape our understanding of the universe and our place within it.

The Importance of High-Energy Astrophysics in Understanding the Universe

Astrophysics, the branch of astronomy that studies the physical properties and behavior of celestial objects, has made remarkable progress in unraveling the mysteries of the universe. However, it is the field of high-energy astrophysics that has truly revolutionized our understanding of the cosmos. In this subchapter, we will explore the importance of high-energy astrophysics and its significant contributions to our knowledge of the universe.

High-energy astrophysics focuses on the study of celestial phenomena that emit or interact with high-energy radiation, such as X-rays and gamma rays. These energetic phenomena include black holes, neutron stars, supernovae, quasars, and gamma-ray bursts. By investigating these extreme cosmic events, scientists have gained unprecedented insights into the fundamental principles governing the universe.

One of the key reasons high-energy astrophysics is vital lies in the fact that the most energetic events in the universe often involve extreme physical conditions that cannot be replicated in terrestrial laboratories. By studying these cosmic events, astrophysicists can test and refine our understanding of physics under extreme conditions, pushing the boundaries of our knowledge.

Furthermore, high-energy astrophysics plays a crucial role in unraveling the mysteries of black holes. These enigmatic cosmic entities, with their immense gravitational pull, have long captivated scientists and the public alike. Through high-energy observations, scientists have been able to study the accretion disks and jets surrounding black holes, shedding light on how matter behaves under the influence of intense gravity.

Additionally, high-energy astrophysics has allowed us to study the early universe. By observing distant quasars, which are highly luminous galactic nuclei powered by accretion onto supermassive black holes, scientists can probe the conditions and processes that prevailed billions of years ago. This provides vital clues about the evolution and formation of galaxies, helping us piece together the cosmic timeline.

Moreover, high-energy astrophysics has enabled the discovery and understanding of gamma-ray bursts (GRBs), the most energetic explosions in the universe. These intense bursts of gamma rays, often associated with the collapse of massive stars, provide a unique window into the physics of extreme energy release and the formation of black holes.

In conclusion, high-energy astrophysics is of paramount importance in understanding the universe. Through studying extreme cosmic events and their associated high-energy radiation, scientists can test the laws of physics under extreme conditions, unravel the mysteries of black holes, explore the early universe, and gain insights into the formation and evolution of galaxies. High-energy astrophysics has truly revolutionized our understanding of the cosmos and will continue to shape our knowledge of the universe for years to come.

Chapter 2: The Electromagnetic Spectrum

Overview of the Electromagnetic Spectrum

The electromagnetic spectrum is a fundamental concept in astrophysics, providing a framework for understanding the diverse range of radiation emitted by celestial objects. From radio waves with the longest wavelengths to gamma rays with the shortest, this spectrum encompasses a wide range of electromagnetic radiation. In this subchapter, we will explore the various regions of the electromagnetic spectrum and their significance in the field of high-energy astrophysics.

The electromagnetic spectrum is crucial in studying astrophysical phenomena because different types of radiation convey unique information about the objects emitting them. The longest wavelengths on the spectrum are occupied by radio waves. These waves are used extensively in radio astronomy to study cosmic phenomena such as pulsars and distant galaxies. By analyzing the radio waves emitted by these objects, astronomers can uncover valuable insights into the structure, composition, and dynamics of the universe.

Moving towards shorter wavelengths, we encounter microwaves and infrared radiation. Microwaves are often associated with heat, and their detection allows scientists to study thermal emissions from objects like interstellar dust clouds or the cosmic microwave background radiation, a remnant of the Big Bang. Infrared radiation, on the other hand, reveals the presence of warm objects in space, including protoplanetary disks and regions of star formation.

Continuing along the spectrum, we reach the region of visible light. This is the range of wavelengths that our eyes are sensitive to, and it has been essential in shaping our understanding of the universe. Visible light observations have led to numerous breakthroughs, from identifying exoplanets to mapping the distribution of dark matter in galaxy clusters.

Beyond visible light, we encounter ultraviolet radiation, X-rays, and finally, gamma rays. These higher-energy forms of electromagnetic radiation are produced by some of the most extreme objects in the universe, such as black holes, supernovae, and gamma-ray bursts. Observing these energetic emissions provides astrophysicists with crucial data to study the physics of these phenomena, as well as the nature of space-time itself.

In conclusion, the electromagnetic spectrum is a vital tool for astrophysicists to unravel the mysteries of high-energy astrophysics. By studying the diverse range of radiation emitted by celestial objects, scientists can gain insights into the nature of the universe, its composition, and the physical processes occurring within it. From radio waves to gamma rays, each region of the electromagnetic spectrum offers a unique window into the cosmos, allowing us to explore the universe in ways that were once unimaginable.

High-Energy Regions of the Electromagnetic Spectrum

The electromagnetic spectrum is a vast range of electromagnetic radiation that encompasses everything from radio waves to gamma rays. Within this spectrum, there are specific regions that are characterized by their high-energy nature. These high-energy regions hold profound mysteries and have captivated astrophysicists for decades. In this subchapter, we will explore these intriguing regions and delve into the mysteries they hold.

One of the most fascinating high-energy regions is the X-ray region. X-rays have a much higher energy than visible light, which allows them to penetrate matter more effectively. This property makes X-rays an invaluable tool for studying objects that are otherwise obscured, such as black holes and neutron stars. X-ray telescopes, like the Chandra X-ray Observatory, have provided astrophysicists with unprecedented insights into the violent and energetic processes occurring in the universe.

Moving further up the energy scale, we encounter the gamma-ray region. Gamma rays are the most energetic form of electromagnetic radiation, often produced in extreme cosmic phenomena like supernovae or active galactic nuclei. Detecting and studying gamma rays poses significant challenges due to their high energy and the Earth's atmosphere blocking most of them. However, space-based observatories like the Fermi Gamma-ray Space Telescope have revolutionized our understanding of these high-energy phenomena.

Within the gamma-ray region, we find another intriguing class of cosmic objects: gamma-ray bursts (GRBs). GRBs are brief, intense bursts of gamma-ray radiation that last from a fraction of a second to a few minutes. They are among the most energetic events in the universe, with energy releases surpassing that of billions of stars combined. Understanding the origin and mechanisms behind GRBs is a major focus of high-energy astrophysics research, and their study has provided invaluable insights into the early universe.

In conclusion, the high-energy regions of the electromagnetic spectrum offer a window into some of the most extreme and mysterious phenomena in the universe. From X-rays to gamma rays, these energetic wavelengths allow us to study black holes, neutron stars, and gamma-ray bursts, providing us with crucial insights into the fundamental workings of our universe. As our understanding of high-energy astrophysics continues to grow, so does our fascination with the awe-inspiring wonders that exist in the cosmos.

Instruments and Techniques for Observing High-Energy Radiation

High-energy astrophysics is a fascinating field that deals with the study of extreme cosmic phenomena such as quasars and gamma-ray bursts. These energetic events emit powerful radiation that can provide us with valuable insights into the nature of the universe. However, observing and studying high-energy radiation is no easy task. It requires specialized instruments and techniques that can handle the immense energy involved and provide accurate data for analysis.

One of the primary instruments used in high-energy astrophysics is the gamma-ray telescope. Unlike optical telescopes that rely on mirrors and lenses to collect and focus visible light, gamma-ray telescopes use detectors to capture high-energy photons. These detectors are often made of scintillation crystals or gas-filled chambers that can register the impact of gamma rays. By precisely measuring the energy and direction of these photons, scientists can determine the source and properties of high-energy radiation.

Another crucial instrument used in this field is the X-ray telescope. X-rays are a form of high-energy radiation that can penetrate matter, making them ideal for studying objects such as black holes and neutron stars. X-ray telescopes employ mirrors with specially designed coatings that can reflect and focus X-rays onto a detector. These telescopes are often placed in space to avoid the absorption of X-rays by the Earth's atmosphere, which is opaque to this type of radiation.

To complement these instruments, astrophysicists also employ various techniques for observing and analyzing high-energy radiation. One commonly used technique is called spectroscopy, which involves breaking down the incoming radiation into its constituent wavelengths or energies. By studying the resulting spectrum, scientists can decipher the chemical composition, temperature, and velocity of the source, providing vital clues about its nature and origin.

Another technique used in high-energy astrophysics is time-resolved imaging. This method involves capturing rapid changes in the intensity or position of a high-energy source over time. By analyzing these variations, scientists can infer the physical processes occurring within the source and gain deeper insights into the underlying astrophysical phenomena.

In conclusion, the study of high-energy radiation in astrophysics requires specialized instruments and techniques tailored to the extreme nature of these cosmic events. Gamma-ray and X-ray telescopes, along with spectroscopy and time-resolved imaging, are invaluable tools in unraveling the mysteries of high-energy astrophysics. Through their usage, scientists can continue to push the boundaries of knowledge and deepen our understanding of the universe we inhabit. Whether you are an astrophysics enthusiast or simply curious about the wonders of the cosmos, exploring the instruments and techniques used in observing high-energy radiation will undoubtedly enhance your appreciation of the vast and complex universe we are a part of.

Chapter 3: Quasars: The Powerhouses of the Universe

What are Quasars?

Quasars, short for quasi-stellar radio sources, are among the most enigmatic and intriguing objects in the known universe. These celestial entities emit vast amounts of energy, making them some of the brightest objects ever observed. In this subchapter, we will dive into the fascinating world of quasars, exploring their origins, properties, and the mysteries they hold.

Quasars were first discovered in the 1960s as mysterious radio sources that appeared star-like in optical observations. However, further investigation revealed that they were anything but stars. Quasars are actually powered by supermassive black holes, millions or even billions of times more massive than our Sun, residing at the centers of galaxies.

The intense energy emitted by quasars is a result of the gravitational forces acting on matter falling into these massive black holes. As matter spirals into the black hole's event horizon, it forms an accretion disk, releasing an enormous amount of energy in the process. This energy is emitted in the form of powerful jets, which can extend thousands of light-years into space.

One of the most intriguing aspects of quasars is their extreme distance from Earth. Due to their immense brightness, quasars can be observed at cosmological distances, meaning they provide valuable insights into the early universe. Studying quasars allows us to probe the universe's history, unravel the formation and evolution of galaxies, and understand the conditions in the early stages of the universe.

Quasars also play a crucial role in our understanding of astrophysics. By studying the absorption lines in their spectra, scientists can gain insights into the intergalactic medium and the distribution of matter in the universe. Quasars also serve as cosmic lighthouses, enabling us to study the properties of space itself, such as the expansion of the universe and the nature of dark matter and dark energy.

In conclusion, quasars are cosmic powerhouses, fueled by supermassive black holes and emitting vast amounts of energy. Their immense brightness and distance make them invaluable tools for studying the cosmos, shedding light on the universe's history and providing crucial insights into astrophysical processes. As we continue to explore the mysteries of high-energy astrophysics, quasars will undoubtedly remain at the forefront of our scientific endeavors.

The Origin and Nature of Quasars

Quasars, short for quasi-stellar radio sources, are enigmatic and extraordinary objects that have captivated the attention of astrophysicists for decades. In this subchapter, we will delve into the fascinating world of quasars, exploring their origin and nature, and shedding light on the mysteries that surround them.

Quasars were first discovered in the 1960s as powerful radio sources with peculiar optical properties. They appear as extremely bright, star-like objects, but their spectra exhibit highly redshifted emission lines. This redshift phenomenon points to the fact that quasars are located at vast distances from Earth, often billions of light-years away. In fact, some of the most distant quasars observed are believed to have existed when the universe was less than a billion years old.

But what exactly are quasars? The prevailing theory suggests that they are powered by supermassive black holes at the centers of galaxies. These black holes, with masses millions or even billions of times greater than our sun, accrete enormous amounts of gas and dust from their surrounding environments. As this matter falls into the black hole's gravitational grip, it forms a swirling, hot disk known as an accretion disk. The intense gravitational forces in the vicinity of the black hole cause the material in the accretion disk to heat up to incredibly high temperatures, emitting vast amounts of energy across the electromagnetic spectrum.

This prodigious release of energy is what makes quasars so luminous. In fact, quasars are among the brightest objects in the universe, outshining entire galaxies. The energy they emit can exceed that of an entire galaxy containing billions of stars. Such immense power output allows quasars to influence their host galaxies, shaping their evolution and triggering the formation of new stars.

Despite extensive research, several aspects of quasars remain mysterious. For instance, scientists are still uncertain about the exact mechanisms that enable the black holes to accrete matter so efficiently. They also seek to understand why only a fraction of galaxies harbor quasars and what triggers their activation. Furthermore, the ultimate fate of quasars, once their fuel is depleted, remains a subject of ongoing investigation.

In conclusion, quasars represent one of the most captivating and enigmatic phenomena in astrophysics. Their immense power, mysterious origins, and profound influence on their host galaxies continue to fascinate scientists and push the boundaries of our understanding of the universe. Through ongoing research and technological advancements, we hope to unravel the secrets of quasars, shedding light on the fundamental processes that shape the cosmos.

Observational Signatures of Quasars

Quasars, or quasi-stellar radio sources, have long captivated the imagination of scientists and enthusiasts alike. These enigmatic celestial objects, powered by supermassive black holes at the centers of distant galaxies, emit tremendous amounts of energy across the electromagnetic spectrum. In this subchapter, we will explore the observational signatures of quasars, shedding light on their nature and offering a glimpse into the mysteries of high-energy astrophysics.

One of the most distinctive features of quasars is their exceptional brightness. Despite their incredible distance from Earth, some quasars outshine entire galaxies, emitting thousands of times more energy than our Milky Way. This luminosity can be observed across a wide range of wavelengths, from radio waves to gamma rays. By studying the spectral energy distribution, astronomers can gain insights into the physical processes occurring within these cosmic powerhouses.

In the radio regime, quasars exhibit a characteristic signature known as core-jet structures. These structures arise from the relativistic jets of charged particles accelerated near the black hole's event horizon. By observing the orientation and motion of these jets, scientists can infer the orientation and spin of the central black hole, providing valuable clues about the dynamics of these systems.

Moving into the optical and ultraviolet regions, quasars display unique emission lines in their spectra. These lines arise from the intense radiation emitted by gas clouds surrounding the central black hole. By analyzing these spectral lines, astronomers can measure the velocity and composition of the gas, helping them understand the structure and evolution of the quasar's immediate environment.

In addition to their optical signatures, quasars also emit X-rays and gamma rays. These high-energy emissions provide crucial information about the accretion process, where matter falls into the black hole. X-ray observations allow us to probe the extreme conditions near the event horizon, providing insights into the physical properties of matter under these extreme gravitational forces.

Studying the observational signatures of quasars not only deepens our understanding of these cosmic powerhouses but also sheds light on broader astrophysical phenomena. Quasars are thought to play a vital role in galaxy evolution, influencing star formation and the growth of galaxies. Moreover, their high-energy emissions can help us explore the properties of the intergalactic medium and the early universe.

In conclusion, the study of quasars offers a fascinating window into the mysteries of high-energy astrophysics. By observing their unique signatures across the electromagnetic spectrum, scientists can uncover the physical processes occurring near supermassive black holes and gain insights into the evolution and dynamics of galaxies. Quasars truly represent one of the most intriguing phenomena in the cosmos, captivating audiences of all backgrounds and leaving us in awe of the wonders of the universe.

The Role of Quasars in Galaxy Formation and Evolution

Quasars, short for quasi-stellar radio sources, are among the most fascinating and mysterious objects in the universe. These enigmatic celestial bodies emit an enormous amount of energy, outshining entire galaxies. In this subchapter, we will explore the crucial role that quasars play in the formation and evolution of galaxies, shedding light on the mysteries of high-energy astrophysics.

Quasars are thought to be powered by supermassive black holes at the centers of galaxies. As matter falls into these black holes, it forms an accretion disk, releasing an immense amount of energy in the process. This energy is emitted in various forms, including visible light, X-rays, and radio waves. The intense luminosity of quasars allows us to study them in great detail, providing valuable insights into the processes occurring in their host galaxies.

One of the fundamental contributions of quasars to galaxy formation is through their role in triggering star formation. The energy released by quasars can compress nearby gas clouds, causing them to collapse and form new stars. This process, known as "quasar feedback," not only influences the growth of galaxies but also regulates their star formation rates. Quasar feedback is crucial in understanding the distribution of stars within galaxies and the overall structure of the universe.

Furthermore, quasars have a profound impact on galaxy evolution through their ability to expel gas from their host galaxies. The intense radiation emitted by quasars can heat and ionize the surrounding gas, driving it away from the galaxy. This process, known as "quasar-driven winds," can significantly affect the gas reservoirs available for future star formation, thus shaping the long-term evolution of galaxies.

Studying quasars also allows us to investigate the early universe. The most distant quasars observed to date provide a window into the universe's infancy, as their light has traveled billions of years to reach us. By analyzing the spectra of these distant quasars, astrophysicists can study the composition and properties of the early universe, providing valuable insights into the formation of the first galaxies.

In conclusion, quasars play a pivotal role in galaxy formation and evolution. Their immense energy output influences star formation rates, triggers the birth of new stars, and expels gas from their host galaxies. The study of quasars allows us to unravel the mysteries of high-energy astrophysics and provides a glimpse into the early stages of the universe. Understanding the role of quasars is crucial for advancing our knowledge of astrophysics and the universe we inhabit.

Chapter 4: Active Galactic Nuclei

Understanding Active Galactic Nuclei

The study of active galactic nuclei (AGN) is a fascinating field within astrophysics that holds the key to unraveling the mysteries of high-energy astrophysics. AGN are the brightest and most energetic objects in the universe, emitting enormous amounts of radiation across the entire electromagnetic spectrum. This subchapter aims to provide a comprehensive overview of AGN, their nature, and their significance in our understanding of the cosmos.

At the heart of every galaxy lies a supermassive black hole, millions to billions of times more massive than our Sun. When matter falls into these black holes, it forms an accretion disk, a swirling mass of hot gas and dust that spirals inward due to gravity. The release of gravitational potential energy as this matter falls towards the black hole is what generates the tremendous amounts of energy observed in AGN.

AGN are classified into different types based on their observational properties. The most common type is known as a quasar, which is characterized by its extreme luminosity and ability to outshine the entire galaxy it resides in. Other types include Seyfert galaxies, blazars, and radio galaxies, each exhibiting unique characteristics that provide valuable insights into the physical processes occurring within AGN.

Studying AGN is crucial in understanding galaxy evolution and the overall structure of the universe. AGN are believed to play a vital role in regulating the growth of galaxies by expelling vast amounts of energy in the form of powerful jets and winds. These jets can extend for thousands of light-years and have a profound impact on the surrounding interstellar medium.

Furthermore, AGN are also thought to be responsible for some of the most energetic phenomena in the universe, such as gamma-ray bursts and cosmic rays. By studying AGN, astrophysicists can gain insights into the mechanisms behind these extreme events and their effects on the surrounding environment.

In conclusion, active galactic nuclei are enigmatic objects that hold the key to understanding the mysteries of high-energy astrophysics. Their immense energy output, diverse properties, and influence on galaxy evolution make them a captivating subject of study. By delving into the intricacies of AGN, scientists can uncover the secrets of the universe and expand our knowledge of the cosmos.

Different Types of Active Galactic Nuclei

In the vast expanse of the cosmos, there exist celestial objects that emit an extraordinary amount of energy, captivating the attention of astrophysicists and stargazers alike. These energetic powerhouses, known as active galactic nuclei (AGN), are found at the centers of galaxies and radiate immense amounts of radiation across the electromagnetic spectrum. Their study has revolutionized our understanding of high-energy astrophysics and continues to unravel the mysteries of the universe.

AGN come in various types, each characterized by unique features and behaviors. Let us explore some of the different types of AGN that have baffled scientists and inspired awe among astronomy enthusiasts.

One of the most famous types is the quasar, short for "quasi-stellar radio source." Quasars represent a phase in the evolution of galaxies where the central supermassive black hole accretes enormous amounts of matter and releases tremendous amounts of energy. Emitting light that can outshine an entire galaxy, quasars are among the brightest objects in the universe. They have played a crucial role in shaping our understanding of cosmology and the early universe.

Another type of AGN is the Seyfert galaxy, named after the American astronomer Carl Seyfert who first identified their distinctive properties. Seyfert galaxies also house supermassive black holes, but they exhibit less extreme activity compared to quasars. They emit strong, characteristic emission lines from their centers, providing insight into the dynamics of the surrounding gas and the black hole's influence on its environment.

Blazars are a peculiar type of AGN that emit intense, highly variable radiation, particularly in the form of gamma rays. They are characterized by strong radio emission and the presence of jets of charged particles traveling at nearly the speed of light. Blazars are believed to be oriented in such a way that their jets are directly aligned with Earth, making them appear exceptionally bright and allowing us to study the physics of particle acceleration and high-energy processes.

These are just a few examples of the diverse array of AGN that populate the cosmos. Studying these energetic phenomena provides astrophysicists with valuable insights into the processes occurring within galaxies, the nature of supermassive black holes, and the evolution of the universe itself. As we delve deeper into the mysteries of high-energy astrophysics, the exploration of AGN continues to captivate scientists and awe-inspired observers alike, offering a glimpse into the extraordinary forces shaping our cosmic neighborhood.

Emission Mechanisms in Active Galactic Nuclei

In the vast expanse of the universe, some of the most awe-inspiring and enigmatic phenomena occur within the hearts of galaxies. Active Galactic Nuclei (AGN) are the powerhouses of the cosmos, emitting colossal amounts of energy across the entire electromagnetic spectrum. Understanding the mechanisms behind this extraordinary emission is essential to unraveling the mysteries of high-energy astrophysics.

AGN are believed to harbor supermassive black holes at their centers, surrounded by a swirling disk of hot gas and dust known as an accretion disk. As matter spirals towards the black hole, immense amounts of gravitational potential energy are released, generating a tremendous luminosity. However, this simple explanation only scratches the surface of the complex and diverse emission mechanisms observed in AGN.

One of the primary mechanisms responsible for the emission in AGN is known as synchrotron radiation. Charged particles, such as electrons, within the accretion disk are accelerated to relativistic speeds by magnetic fields. As these particles spiral along the magnetic field lines, they emit synchrotron radiation across a wide range of frequencies, from radio waves to X-rays. This mechanism is particularly important in the production of powerful radio jets, which can extend thousands of light-years from the central black hole.

Another emission mechanism observed in AGN is inverse Compton scattering. High-energy electrons within the accretion disk can collide with low-energy photons, boosting their energy to X-ray or gamma-ray levels. This process, known as inverse Compton scattering, is responsible for the production of the most energetic radiation emitted by AGN. Understanding the interplay between synchrotron radiation and inverse Compton scattering is crucial to comprehending the overall energy output of AGN.

Moreover, the presence of broad emission lines in the optical and ultraviolet spectra of AGN indicates the presence of a powerful outflowing wind of ionized gas. These emission lines are produced when energetic photons from the accretion disk ionize the surrounding gas, causing it to emit light at specific wavelengths. The study of these emission lines provides valuable insights into the physical conditions and dynamics of the gas near the black hole.

The emission mechanisms in AGN are a fascinating and complex topic within the realm of astrophysics. By unraveling the mysteries surrounding these mechanisms, scientists can gain a deeper understanding of the fundamental processes occurring in the most extreme environments in the universe. Through ongoing research and technological advancements, we continue to explore the enigmatic nature of AGN, shedding light on the mysteries of high-energy astrophysics and expanding our knowledge of the cosmos.

The Connection between Active Galactic Nuclei and Quasars

In the vastness of the cosmos, there exist celestial objects that emit mind-boggling amounts of energy, captivating astronomers and astrophysics enthusiasts alike. Among these enigmatic entities are active galactic nuclei (AGN) and quasars. This subchapter aims to shed light on the profound connection between these two cosmic phenomena, delving into the mysteries of high-energy astrophysics.

Active galactic nuclei are found at the centers of galaxies, where a supermassive black hole resides. These colossal black holes, with masses millions to billions of times that of our sun, have a profound influence on their host galaxies. As matter spirals into the black hole's gravitational grasp, it forms an accretion disk, releasing staggering amounts of energy in the process. This intense emission is what characterizes AGN.

Quasars, short for quasi-stellar radio sources, are a subset of AGN that exhibit an extraordinarily high luminosity. These cosmic powerhouses outshine entire galaxies, their brilliance emanating from the immense energy released as matter falls into the supermassive black hole at the AGN's core. The term "quasar" was coined due to their star-like appearance when observed from Earth.

The connection between AGN and quasars lies in their shared origins. Quasars are believed to be the most luminous stage in the life cycle of an AGN. As matter accretes onto the black hole, the release of energy causes the AGN to transition into a quasar state, reaching its peak luminosity. However, not all AGN evolve into quasars, making these rare objects a subject of great fascination.

Studying the connection between AGN and quasars provides invaluable insights into the formation and evolution of galaxies. By exploring the mechanisms that trigger the transition from AGN to quasar states, astrophysicists gain a deeper understanding of the processes governing galaxy growth. Additionally, quasars serve as beacons that enable astronomers to probe the early universe, as their tremendous luminosity allows their detection even at extreme distances.

For everyone captivated by the wonders of astrophysics, comprehending the connection between AGN and quasars is a gateway to unraveling the mysteries of our universe. The exploration of these cosmic powerhouses not only expands our knowledge of high-energy astrophysics but also provides a glimpse into the origins and evolution of galaxies. As our understanding deepens, so too does our appreciation for the immense beauty and complexity of the cosmos.

Chapter 5: Gamma-Ray Bursts: Cosmic Explosions

The Discovery of Gamma-Ray Bursts

Gamma-ray bursts (GRBs) are some of the most powerful and mysterious phenomena in the universe. They were first discovered in the late 1960s during the early days of space exploration. This subchapter explores the fascinating story behind the discovery of these intense bursts of gamma-ray radiation and their subsequent study in the field of astrophysics.

The story begins with the Vela satellites, a series of satellites launched by the United States to monitor for nuclear weapons tests in space. These satellites were equipped with gamma-ray detectors, and it was during their mission that the first gamma-ray bursts were detected. Initially, these bursts were thought to be the result of nuclear explosions on Earth, but further investigation revealed that they were actually coming from deep space.

As astronomers and scientists began to study these bursts more closely, they realized that they were unlike any other known astrophysical phenomenon. The bursts were incredibly short-lived, lasting only a few seconds to a few minutes, but during that time, they released more energy than the Sun would emit in its entire 10-billion-year lifetime. This made them one of the most energetic events in the universe.

In the years following the discovery, astronomers started to observe these bursts with more sophisticated instruments and telescopes. They found that the bursts were not evenly distributed across the sky, but rather appeared randomly and unpredictably. This raised even more questions about their origin and nature.

Several theories were proposed to explain the origins of gamma-ray bursts. One popular idea was that they were the result of supernovae, the explosive deaths of massive stars. Another theory suggested that they were the result of the merging of neutron stars or black holes. It wasn't until the late 1990s that scientists discovered that there were two distinct types of gamma-ray bursts: long duration bursts, associated with the deaths of massive stars, and short duration bursts, linked to the mergers of compact objects.

The study of gamma-ray bursts continues to be an active area of research in astrophysics. Scientists are using advanced telescopes and satellites to better understand the physical processes involved in these intense events and to shed light on the nature of the universe itself. The discovery of gamma-ray bursts has opened up a new window into the high-energy universe, allowing us to unravel some of its most intriguing mysteries.

In conclusion, the discovery of gamma-ray bursts has revolutionized our understanding of the universe. This subchapter has provided a glimpse into the fascinating story behind their discovery and the subsequent research conducted in the field of astrophysics. Gamma-ray bursts continue to captivate scientists and astrophysics enthusiasts alike, as we strive to uncover the secrets of these powerful cosmic events.

The Nature and Origin of Gamma-Ray Bursts

Gamma-ray bursts (GRBs) are among the most intriguing and enigmatic phenomena in the universe. They are intense, short-lived bursts of high-energy gamma rays that can last from a few milliseconds to several minutes. These cosmic explosions emit more energy in a few seconds than the Sun will radiate in its entire lifetime. Understanding the nature and origin of GRBs is a crucial endeavor in the field of astrophysics, as they offer unique insights into the extreme physics and the early universe.

GRBs were first discovered in the late 1960s by military satellites designed to detect nuclear weapon tests. It wasn't until the 1990s that astronomers were able to pinpoint their celestial origins, thanks to advances in satellite technology. These bursts occur randomly and appear in all parts of the sky, making them challenging to study. Nevertheless, their incredible brightness and detectable afterglows have helped astronomers gather significant information about their properties.

There are two main types of GRBs: long-duration and short-duration bursts. Long-duration GRBs are typically associated with the explosive deaths of massive stars, known as supernovae. These occur in star-forming regions of galaxies and are often found in spiral galaxies. On the other hand, short-duration GRBs are believed to result from the merger of compact objects, such as neutron stars or black holes. These mergers occur in regions of the universe where star formation is less active, such as elliptical galaxies.

The exact mechanisms behind GRBs are still being actively investigated, but it is widely accepted that they involve the release of an enormous amount of energy during cataclysmic events. The leading theory for long-duration GRBs involves the collapse of a massive star's core, forming a black hole. As matter falls into the black hole, powerful jets of particles and radiation are launched along its rotational axis, producing the observed burst of gamma rays.

Short-duration GRBs, on the other hand, are thought to result from the merger of two compact objects. As these objects spiral towards each other, they emit gravitational waves that carry away energy and angular momentum. The final merger releases a tremendous amount of energy, leading to a gamma-ray burst.

Studying GRBs provides invaluable information about the early universe, as they are detectable from extreme distances. By analyzing the afterglows and the spectra of GRBs, astrophysicists can probe the composition and evolution of galaxies, as well as the nature of the interstellar medium.

In conclusion, gamma-ray bursts are awe-inspiring cosmic explosions that continue to captivate astrophysicists and astronomers. Their mysterious nature and immense energy offer a unique window into the extreme physics of the universe. By unraveling the mysteries behind GRBs, we can deepen our understanding of the early universe and the fundamental processes that shape it.

Gamma-Ray Burst Classification and Characteristics

Gamma-ray bursts (GRBs) are some of the most powerful and enigmatic phenomena in the universe. They are brief but intense bursts of gamma-ray radiation that can last from a few milliseconds to several minutes. Discovered in the late 1960s, GRBs have since captivated the attention of scientists and astrophysicists, pushing the boundaries of our understanding of high-energy astrophysics.

In this subchapter, we will explore the classification and characteristics of gamma-ray bursts. GRBs are generally categorized into two classes: long-duration bursts and short-duration bursts. Long-duration bursts last for more than two seconds and are typically associated with the collapse of massive stars, known as supernovae. Short-duration bursts, on the other hand, last for less than two seconds and are believed to originate from the merger of compact objects, such as neutron stars or black holes.

The characteristics of gamma-ray bursts are truly remarkable. They release an immense amount of energy, equivalent to the conversion of several solar masses into pure energy in just a few seconds. This energy is emitted in the form of gamma rays, which are the most energetic form of electromagnetic radiation. The peak energy of the gamma rays can range from a few kiloelectron volts (keV) to several hundreds of GeV (gigaelectron volts).

GRBs can also emit radiation across the entire electromagnetic spectrum, from gamma rays to X-rays, ultraviolet, optical, infrared, and even radio waves. This multi-wavelength emission allows astronomers to study the properties of GRBs and their host galaxies.

Further classification of GRBs is based on their spectral properties, which can be analyzed by studying the different wavelengths of light emitted during the burst. The majority of GRBs exhibit a characteristic spectral shape known as a "Band function," which has a peak energy and a low-energy spectral index. This spectral information provides insights into the physics of the emission mechanism and the nature of the sources.

Understanding the classification and characteristics of gamma-ray bursts is crucial for unraveling the mysteries of high-energy astrophysics. It allows scientists to study the extreme physical processes involved in the birth and death of massive stars, as well as the dynamics of compact object mergers. By observing and analyzing these cosmic explosions, we can gain valuable insights into the nature of the early universe, the formation of black holes, and the production of heavy elements.

In conclusion, gamma-ray bursts are fascinating cosmic phenomena that continue to astound and challenge our understanding of the universe. Their classification and characteristics provide a framework for studying their origins and properties. Through ongoing research and technological advancements, astrophysicists strive to decipher the secrets concealed within these powerful bursts of gamma-ray radiation.

Probing the High-Energy Universe through Gamma-Ray Bursts

Gamma-ray bursts (GRBs) are among the most fascinating and enigmatic events in the universe. These intense and brief bursts of gamma-ray radiation have puzzled astrophysicists for decades, and their study has greatly contributed to our understanding of the high-energy universe. In this subchapter, we will delve into the captivating world of GRBs and explore how they are helping unravel the mysteries of high-energy astrophysics.

To comprehend the significance of GRBs, we must first grasp the sheer power they possess. These cosmic explosions release an astonishing amount of energy in just a few seconds, equivalent to the energy emitted by our Sun over its entire lifespan. The origins of GRBs remained elusive until the late 20th century when breakthrough observations began shedding light on their nature.

GRBs are believed to originate from the cataclysmic deaths of massive stars or the mergers of compact objects such as neutron stars or black holes. When these events occur, a tremendous amount of energy is released in the form of gamma rays. However, due to their high energy and short duration, detecting and studying GRBs is no easy task.

Fortunately, advancements in space-based observatories and ground-based detectors have allowed scientists to capture and study these elusive phenomena. The subchapter will explore the instruments and techniques used to detect GRBs and how they provide valuable insights into the physics of the high-energy universe.

Moreover, we will delve into the various aspects of GRBs, including their classification, duration, and spectral properties. By analyzing these characteristics, scientists have been able to distinguish between different types of GRBs and gain a deeper understanding of their underlying mechanisms.

Furthermore, we will discuss the role GRBs play in cosmology and the study of the early universe. GRBs serve as cosmic beacons, enabling astronomers to probe the most distant and ancient regions of the universe. By examining the light emitted from these explosions, scientists can infer valuable information about the early stages of the universe's evolution.

Lastly, we will explore the current and future missions dedicated to studying GRBs, such as NASA's Swift and Fermi telescopes and the upcoming European Space Agency's Athena mission. These missions promise to revolutionize our understanding of GRBs and their role in shaping the high-energy universe.

In conclusion, this subchapter will provide a comprehensive overview of GRBs, from their origins and detection to their role in astrophysics and cosmology. Whether you are an astrophysics enthusiast or simply curious about the mysteries of the universe, this subchapter will offer a captivating journey through the high-energy universe of gamma-ray bursts.

Chapter 6: High-Energy Jets and Relativistic Outflows

Understanding High-Energy Jets

High-energy jets are one of the most fascinating and enigmatic phenomena in the realm of astrophysics. These powerful streams of particles, propelled at nearly the speed of light, occur in various cosmic objects, including quasars and gamma-ray bursts. In this subchapter, we will delve into the intricacies of high-energy jets, exploring their formation, properties, and the mysteries that surround them.

To comprehend high-energy jets, we must first understand their origin. These jets are often born in the vicinity of black holes or neutron stars, where immense gravitational forces create intense magnetic fields. As matter spirals towards these compact objects, a fraction of it gets channeled into narrow beams along the magnetic field lines. These beams then accelerate to astonishing speeds, generating the highly energetic jets that captivate astrophysicists.

The properties of high-energy jets are awe-inspiring. Their velocities can reach astronomical values, with some jets traveling at a significant fraction of the speed of light. These jets can extend over vast distances, spanning millions of light-years, and release colossal amounts of energy into the surrounding space. They emit radiation across the electromagnetic spectrum, from radio waves to X-rays, offering valuable insights into the extreme conditions near black holes and other energetic astrophysical systems.

Despite decades of research, high-energy jets remain shrouded in mystery. One puzzling aspect is the mechanism responsible for their incredible acceleration. Scientists have proposed various theories, such as the extraction of rotational energy from black holes or the conversion of magnetic energy into kinetic energy. However, the exact process is still not fully understood, and ongoing investigations aim to unravel this enigma.

Furthermore, the interaction between high-energy jets and their surrounding environment presents another intriguing puzzle. These jets have a profound impact on the galaxies or stellar remnants they inhabit, shaping their evolution and influencing the formation of stars. Understanding these interactions is crucial for comprehending the broader astrophysical processes at play.

In conclusion, the study of high-energy jets offers a window into the extreme and captivating phenomena occurring in the universe. Their formation, properties, and interactions hold the key to unraveling the mysteries of black holes, neutron stars, and other energetic astrophysical objects. By unraveling these enigmas, we inch closer to comprehending the fundamental workings of the cosmos and our place within it. Whether you are a seasoned astrophysicist or simply curious about the wonders of the universe, exploring high-energy jets will undoubtedly ignite your imagination and deepen your understanding of the cosmos.

The Role of Magnetic Fields in Jet Formation

In the vast expanse of the cosmos, there are phenomena that challenge our understanding of the universe. From quasars to gamma-ray bursts, high-energy astrophysics is an area of study that unveils the mysteries of celestial objects that emit enormous amounts of energy. One of the most intriguing aspects of these phenomena is the formation of powerful jets, and magnetic fields play a crucial role in this process.

Jets, in astrophysics, refer to narrow streams of particles that are accelerated to incredible speeds. These jets can span thousands of light-years and release energy equivalent to that of millions of stars. They emanate from a wide variety of celestial objects, such as quasars, black holes, and neutron stars, and their formation has baffled scientists for decades. However, recent research has shed light on the pivotal role played by magnetic fields in the creation and propulsion of these extraordinary jets.

Magnetic fields are pervasive in the universe, and their influence on celestial bodies is undeniable. In the context of jet formation, magnetic fields can become tangled and twisted due to various astrophysical processes. When this happens near a central object, such as a black hole, the magnetic field lines can become tightly wrapped around the object's rotational axis. As matter spirals toward the central object, these twisted magnetic field lines amplify and generate a powerful magnetic field.

The intense magnetic field then interacts with the surrounding matter, causing it to be channeled along the rotational axis in the form of a jet. This phenomenon, known as the Blandford-Znajek process, is responsible for the creation of some of the most energetic jets observed in the universe. The magnetic field acts as a conduit, accelerating particles to relativistic speeds and propelling them along the jet's length.

Understanding the role of magnetic fields in jet formation is crucial for unraveling the mysteries of high-energy astrophysics. By studying the properties of these magnetic fields and their interaction with matter, scientists can gain insights into the physics behind these enigmatic jets. This knowledge not only enhances our understanding of the universe but also provides a foundation for future advancements in astrophysics.

In conclusion, the role of magnetic fields in jet formation is a captivating aspect of high-energy astrophysics. These powerful jets, emanating from celestial objects like black holes and quasars, are fueled by the interaction of intense magnetic fields with surrounding matter. By untangling the mysteries of magnetic fields and their impact on jet formation, scientists are able to deepen our understanding of the universe and uncover the secrets of these awe-inspiring phenomena.

Observations and Properties of High-Energy Jets

In the vast expanse of the universe, there are celestial objects that emit mind-boggling amounts of energy, far beyond what we can comprehend. These phenomena, known as high-energy jets, have captivated the attention of scientists and astrophysicists for decades. In this subchapter, we will delve into the intriguing observations and properties of these enigmatic cosmic jets, aiming to unravel the mysteries of high-energy astrophysics.

High-energy jets are colossal beams of particles and radiation that are propelled at incredible speeds from certain astrophysical objects. They can be found emanating from various sources, including quasars, blazars, and gamma-ray bursts. These jets are thought to be powered by the intense gravitational pull of black holes or the rapid accretion of matter onto compact objects.

One of the remarkable properties of high-energy jets is their astonishing brightness. These jets can outshine entire galaxies, emitting radiation across the entire electromagnetic spectrum, from radio waves to gamma rays. Through advanced telescopes and space-based observatories, scientists have been able to study these jets in great detail, providing vital insights into their nature.

Observations have revealed that high-energy jets are highly collimated, meaning that they maintain a narrow beam of particles over vast distances. This characteristic suggests the presence of a mechanism that focuses and channels the energy in a concentrated manner. Additionally, these jets exhibit extreme variability, with fluctuations in brightness occurring on timescales as short as minutes or even seconds. Such rapid changes challenge our understanding of how these jets are formed and sustained.

Furthermore, high-energy jets often display a phenomenon called relativistic beaming. Due to the motion of the jet towards Earth, the emitted radiation appears more intense, enhancing its apparent brightness. This effect has significant implications for our ability to detect and study these jets, as it influences our interpretation of their properties.

Understanding the origin and behavior of high-energy jets is crucial for unraveling the mysteries of the universe. By studying their emissions, scientists hope to gain insights into the fundamental processes occurring near black holes and other extreme cosmic environments. These observations not only shed light on the astrophysical phenomena themselves but also provide valuable information about the fundamental laws of physics that govern our universe.

In conclusion, the observations and properties of high-energy jets continue to fascinate and challenge scientists in the field of astrophysics. Through advanced telescopes and space-based observatories, we have been able to gain valuable insights into these cosmic phenomena. However, much remains to be discovered, and the study of high-energy jets promises to unveil new secrets about the nature of our universe.

The Implications of Relativistic Outflows in Astrophysics

Relativistic outflows, also known as jets, are fascinating phenomena that play a crucial role in astrophysics. They have been observed in various cosmic objects, from quasars to gamma-ray bursts, and continue to puzzle scientists with their immense energy and high velocities. Understanding the implications of relativistic outflows is essential for unraveling the mysteries of high-energy astrophysics. This subchapter explores the significance of these outflows and their impact on our understanding of the universe.

Relativistic outflows are narrow streams of particles and radiation that are ejected from the central regions of astrophysical objects. These outflows can reach velocities close to the speed of light, making them highly energetic and capable of transporting vast amounts of energy across large distances. They are thought to be powered by the gravitational potential energy of accreting matter onto compact objects, such as black holes or neutron stars.

One of the most profound implications of relativistic outflows is their role in shaping the evolution of galaxies. Jets emitted from active galactic nuclei (AGN) have a significant impact on their surrounding environments. They can deposit enormous amounts of energy into the interstellar medium, affecting star formation, galaxy morphology, and even the growth of supermassive black holes. By studying the properties of AGN jets, astrophysicists can gain insights into the formation and evolution of galaxies throughout cosmic history.

Relativistic outflows are also intimately connected to the production of high-energy radiation. Gamma-ray bursts (GRBs) are among the most energetic events in the universe, and their origin has been linked to the relativistic jets of collapsing massive stars or merging compact objects. By investigating the properties of GRB jets, scientists can unveil the nature of matter in extreme conditions and probe the fundamental laws of physics.

Furthermore, relativistic outflows serve as cosmic laboratories for testing the theory of general relativity. The intense gravitational fields near black holes and neutron stars can cause significant distortions in the jet's trajectory and emission properties. Studying these effects can help refine and validate Einstein's theory, which underpins our understanding of gravity.

In conclusion, relativistic outflows hold great importance in astrophysics and have far-reaching implications for our understanding of the universe. Their influence extends from shaping the evolution of galaxies to shedding light on the mysteries of high-energy radiation and testing fundamental theories of physics. By further investigating these phenomena, scientists can continue to unravel the mysteries of high-energy astrophysics and gain valuable insights into the nature of our cosmos.

Chapter 7: Supernovae and Neutron Stars

Supernovae: Stellar Explosions

In the vast expanse of the universe, there are events of unimaginable power and beauty that shape the cosmos as we know it. Among these cosmic spectacles, supernovae stand out as some of the most awe-inspiring and intriguing phenomena in the field of astrophysics. These stellar explosions, occurring in the final stages of a star's life, are a subject of fascination for scientists and enthusiasts alike.

Supernovae mark the explosive end of massive stars, whose nuclear fuel has been depleted after millions of years of fusion. As gravity takes over, the star's core collapses under its immense weight, triggering a cataclysmic explosion that releases an incredible amount of energy. The resulting blast can outshine an entire galaxy for weeks, and during this brief period, a supernova becomes one of the brightest objects in the universe.

One of the key aspects of supernovae is their role in cosmic evolution. These titanic explosions disperse heavy elements, such as iron and gold, into space, enriching the interstellar medium and providing the building blocks for future generations of stars. Supernovae are also responsible for the creation and distribution of elements necessary for life, including carbon and oxygen. Without these stellar explosions, the universe as we know it would not exist.

Scientists classify supernovae into different types based on their characteristics. Type Ia supernovae occur in binary star systems, where a white dwarf star accretes matter from a companion star until it reaches a critical mass, triggering a runaway nuclear fusion reaction. Type II supernovae, on the other hand, are the result of the core collapse of massive stars. These two types have distinct signatures that allow astronomers to study the physics behind these explosions and gain insight into the nature of the universe.

Understanding supernovae is crucial not only for astrophysics but also for other scientific fields. These cosmic events serve as cosmic laboratories, providing valuable data on extreme physics, including the behavior of matter under extreme densities and temperatures. Moreover, studying supernovae helps us unravel the mysteries of dark energy, which is believed to be responsible for the accelerated expansion of the universe.

In conclusion, supernovae are celestial fireworks that captivate astronomers and laypeople alike. These explosive stellar deaths shape the universe and provide invaluable insights into the workings of the cosmos. From the formation of elements to the expansion of the universe, the study of supernovae has profound implications for our understanding of high-energy astrophysics and the grand tapestry of the universe we call home.

Neutron Stars: Nature's Densest Objects

Neutron stars are among the most fascinating and enigmatic celestial objects in the universe. With densities so extreme that even a teaspoonful of their material would weigh billions of tons, these cosmic wonders continue to captivate the minds of astrophysicists and enthusiasts alike. In this subchapter, we delve into the astounding nature of neutron stars, unraveling the mysteries that surround them and exploring their significance in the field of high-energy astrophysics.

When a massive star exhausts its nuclear fuel, it often undergoes a supernova explosion, leaving behind a dense core. This core, known as a neutron star, is the remnant of a once-mighty star, compressed to a size comparable to a city, yet containing more mass than our Sun. The gravitational pull on a neutron star is so intense that its outer layers collapse inward, resulting in a core composed primarily of neutrons.

One of the most intriguing aspects of neutron stars is their incredible density. To put it into perspective, imagine compressing the entire human population into a space smaller than a sugar cube. This extreme density leads to remarkable properties, such as their ability to generate powerful magnetic fields and emit beams of radiation from their poles. These beams, only visible when they point towards Earth, create a phenomenon known as a pulsar – a rapidly rotating neutron star that appears to pulse with regular intervals.

Furthermore, neutron stars serve as celestial laboratories for studying the fundamental laws of physics. The intense pressures and temperatures within their cores give rise to exotic states of matter, including superfluids and superconductors. By observing the behavior of matter under such extreme conditions, scientists gain insights into the nature of the strong nuclear force and the behavior of subatomic particles.

Neutron stars also play a crucial role in the formation of elements heavier than iron. During a supernova explosion, the incredible heat and pressure generated by the collapsing star trigger a process called nucleosynthesis. This process fuses atomic nuclei, creating elements like gold, platinum, and uranium. Without neutron stars, these precious elements would not exist in the universe.

In conclusion, neutron stars are indeed nature's densest objects, captivating our imagination and pushing the boundaries of our understanding of the cosmos. Through their unique properties and behaviors, they offer invaluable insights into the workings of the universe at its most extreme scales. Whether you are an astrophysics enthusiast or simply curious about the wonders of the cosmos, the study of neutron stars will undoubtedly leave you in awe of the immense diversity and complexity of our universe.

Supernova Remnants and High-Energy Emission

Supernova remnants (SNRs) and high-energy emission are fascinating phenomena that play a crucial role in our understanding of the universe. In this subchapter, we will delve into the mysteries surrounding these cosmic events and their implications in the field of astrophysics.

When a massive star reaches the end of its life, it undergoes a catastrophic explosion known as a supernova. This explosion releases an enormous amount of energy, equal to the energy emitted by our Sun over its entire lifetime. The remnants left behind from these explosions, called SNRs, become cosmic laboratories for studying the high-energy processes occurring in the universe.

One of the most intriguing aspects of SNRs is their ability to emit high-energy radiation, including X-rays and gamma-rays. These energetic emissions provide valuable insights into the physical mechanisms happening within these remnants. Scientists believe that the high-energy emission from SNRs is mainly caused by the interaction between the expanding remnant and the surrounding interstellar medium.

In the subchapter "Supernova Remnants and High-Energy Emission," we will explore the different types of high-energy emission observed from SNRs. We will discuss the mechanisms responsible for producing X-rays and gamma-rays, such as synchrotron radiation, inverse Compton scattering, and non-thermal particle acceleration.

Furthermore, we will delve into the role of SNRs in cosmic ray production. Cosmic rays are high-energy particles, mostly protons and atomic nuclei, that travel through space at nearly the speed of light. SNRs are believed to be the primary sources of these cosmic rays, as the shockwaves generated during the supernova explosion can accelerate charged particles to extreme energies.

In this subchapter, we will also touch upon the significance of studying SNRs and high-energy emission in the broader context of astrophysics. Understanding the mechanisms behind these processes not only sheds light on the life and death of massive stars but also contributes to our knowledge of the evolution of galaxies, the formation of elements, and the interplay between cosmic particles and magnetic fields.

Whether you are an astrophysics enthusiast, a student, or simply curious about the wonders of the universe, "Supernova Remnants and High-Energy Emission" offers an accessible and captivating exploration of these extraordinary cosmic phenomena. Join us as we unravel the mysteries of high-energy astrophysics and dive into the awe-inspiring world of supernova remnants.

Pulsars: Rotating Neutron Stars and Their High-Energy Emission

In the vast expanse of the universe, there exist celestial objects that continue to captivate and mystify astrophysicists and astronomers alike. One such enigmatic entity is the pulsar, a rotating neutron star that emits high-energy radiation. This subchapter delves into the fascinating world of pulsars, unraveling their mysteries and shedding light on their role in high-energy astrophysics.

Neutron stars are the remnants of massive stars that have undergone a supernova explosion. These incredibly dense stellar remnants, with masses greater than that of the Sun but compressed into a sphere just a few kilometers in diameter, possess immense gravitational forces. Pulsars are a specific class of neutron stars that emit beams of electromagnetic radiation, similar to a cosmic lighthouse.

The rotation of a pulsar generates a powerful magnetic field, and as the magnetic poles do not align with the rotational axis, the emitted radiation appears as periodic pulses when viewed from Earth. These pulses can be observed across multiple wavelengths, from radio waves to X-rays and even gamma rays. Pulsars thus offer a unique opportunity to study high-energy astrophysics.

One of the key questions surrounding pulsars is the origin of their high-energy emission. The primary mechanism responsible for the radiation is thought to be synchrotron radiation, a process where high-energy particles spiral along magnetic field lines, emitting radiation in the process. However, other processes such as inverse Compton scattering and curvature radiation also contribute to the high-energy emission observed from pulsars.

Studying pulsars has provided valuable insights into various astrophysical phenomena. For instance, their precise periodicity makes them excellent natural clocks, enabling scientists to test theories of gravity and measure the properties of spacetime. Pulsars also serve as powerful probes of the interstellar medium, allowing scientists to study the properties of the material between stars.

Understanding pulsars is of paramount importance in the field of astrophysics. These peculiar objects not only offer a window into the physics of extreme environments but also provide valuable information about the evolution of stars and galaxies. With ongoing advancements in observational techniques and theoretical modeling, the study of pulsars continues to push the boundaries of our knowledge, unraveling the mysteries of high-energy astrophysics.

Whether you are an astrophysics enthusiast or simply curious about the wonders of the universe, this subchapter on pulsars will take you on an exciting journey through the intricacies of rotating neutron stars. Join us as we explore the fascinating world of pulsars and their high-energy emission, ultimately revealing the secrets of high-energy astrophysics that continue to shape our understanding of the cosmos.

Chapter 8: Black Holes: The Ultimate Powerhouses

Introduction to Black Holes

Subchapter: Introduction to Black Holes

Welcome to the fascinating world of black holes! In this subchapter, we will embark on a journey to unravel the mysteries of these enigmatic cosmic objects. Whether you are an avid enthusiast of astrophysics or someone who has just begun to explore the wonders of the universe, this introduction will provide you with a solid understanding of the captivating phenomenon known as black holes.

Black holes represent one of the most intriguing and mind-boggling concepts in the field of astrophysics. They are regions in space where gravity is so incredibly strong that nothing, not even light, can escape their gravitational pull. Imagine a cosmic vacuum cleaner, devouring everything that comes within its reach. That, in essence, is a black hole.

Our exploration begins with understanding the formation of black holes. When a massive star exhausts its nuclear fuel, it undergoes a catastrophic collapse under its own gravitational force. The core of the star collapses inward, forming a singularity, an infinitely dense point where all the matter is concentrated. Surrounding this singularity is the event horizon, the point of no return beyond which nothing can escape.

As we delve deeper, we will explore the different types of black holes. Stellar black holes are the remnants of massive stars that have undergone gravitational collapse. Supermassive black holes, on the other hand, reside at the centers of galaxies and can be millions or even billions of times more massive than our sun. These cosmic giants have a profound impact on the surrounding environment, shaping the evolution of galaxies.

We will also discuss the fascinating phenomena associated with black holes, such as Hawking radiation, which suggests that black holes can emit particles and slowly lose mass over time. Additionally, we will unravel the mysteries of black hole mergers, where two black holes spiral towards each other, releasing vast amounts of gravitational waves.

While black holes are undoubtedly captivating, they also present numerous unanswered questions. Are there smaller black holes lurking in the depths of space? What lies beyond the event horizon? Can black holes be used as portals to other dimensions? These questions, and many more, continue to challenge our understanding of the universe.

So, whether you are a curious beginner or an astrophysics aficionado, join us as we embark on this thrilling exploration into the realm of black holes. Prepare to be amazed, astounded, and inspired as we unravel the mysteries of these cosmic behemoths and deepen our understanding of the captivating field of astrophysics.

Stellar-Mass Black Holes

Black holes have long remained one of the most enigmatic and captivating entities in the universe. Among the various types of black holes, stellar-mass black holes hold a special place of intrigue for astrophysicists and enthusiasts alike. In this subchapter, we delve into the fascinating world of these celestial objects, unlocking the mysteries surrounding their formation, behavior, and significance.

Stellar-mass black holes are formed through the gravitational collapse of massive stars. When a star exhausts its nuclear fuel, it undergoes a supernova explosion, resulting in the formation of a black hole. These black holes typically have a mass ranging from a few times that of our Sun to several tens of solar masses, hence the name "stellar-mass."

One of the most remarkable aspects of stellar-mass black holes is their gravitational pull, which is so strong that not even light can escape its clutches. This phenomenon is known as the event horizon, a boundary beyond which nothing can escape the black hole's gravitational grasp. The event horizon is a defining characteristic of black holes, and its study has provided invaluable insights into the nature of these cosmic marvels.

Stellar-mass black holes are not only captivating due to their gravitational pull but also because of the extraordinary phenomena that occur in their vicinity. As matter spirals into a black hole, it forms an accretion disk, releasing immense amounts of energy in the form of X-rays and gamma rays. These high-energy emissions can be detected by specialized instruments, revealing the presence and characteristics of stellar-mass black holes.

Furthermore, the study of stellar-mass black holes contributes to our understanding of the evolution of galaxies. Black holes are believed to play a crucial role in the formation and growth of galaxies, as their immense gravitational influence shapes the surrounding environment. By studying the behavior and distribution of stellar-mass black holes, astrophysicists can unravel the intricate connections between black holes, galaxies, and the larger cosmic web.

In conclusion, stellar-mass black holes represent a captivating and important field of study within astrophysics. Their formation, behavior, and influence shed light on the fundamental processes that shape our universe. From their gravitational pull to the high-energy emissions they produce, stellar-mass black holes continue to unravel the mysteries of high-energy astrophysics, captivating the imagination of scientists and enthusiasts alike.

Supermassive Black Holes

In the vast expanse of the universe, there exist celestial objects that capture the imagination and challenge our understanding of the cosmos. Among these enigmatic entities, none are as mind-boggling as supermassive black holes. This subchapter will delve into the depths of these cosmic monsters, exploring their origins, characteristics, and the role they play in shaping our universe.

Supermassive black holes are titanic gravitational beasts residing at the hearts of galaxies, including our own Milky Way. They possess masses millions or even billions of times that of our sun, concentrated within a region so small that it defies comprehension. These cosmic behemoths are formed through a complex interplay of galactic dynamics and the relentless pull of gravity.

Despite their name, black holes are not holes in the traditional sense. Rather, they are incredibly dense objects born from the collapse of massive stars. When a star exhausts its nuclear fuel, it undergoes a cataclysmic explosion known as a supernova. If the core of the star is sufficiently massive, it collapses under its own gravity, forming a black hole. However, the exact mechanisms behind the formation of supermassive black holes are still a subject of intense scientific investigation.

Once formed, these cosmic leviathans continue to grow by devouring matter in their vicinity. As gas and dust fall into their gravitational clutches, it forms an accretion disk around the black hole. This disk can reach temperatures of millions of degrees, emitting tremendous amounts of energy in the form of X-rays and gamma rays. These energetic emissions make supermassive black holes visible to telescopes, allowing scientists to study their properties.

However, the impact of supermassive black holes extends far beyond their immediate surroundings. Their immense gravitational pull can influence the evolution of entire galaxies. As matter falls into the black hole, it releases powerful jets of particles that can shape the formation of stars and influence the dynamics of their host galaxies. Understanding these interactions is crucial for comprehending the intricate dance between black holes and galaxies throughout cosmic history.

The study of supermassive black holes is a rapidly advancing field within astrophysics. Scientists are employing increasingly sophisticated techniques, such as gravitational wave detectors and high-resolution imaging, to unravel the mysteries of these cosmic giants. As our knowledge grows, so too does our understanding of the universe and our place within it.

In conclusion, supermassive black holes are captivating objects that challenge our understanding of the cosmos. From their mysterious origins to their profound influence on galactic evolution, these celestial monsters continue to fascinate scientists and astrophysics enthusiasts alike. By exploring these immense cosmic phenomena, we come one step closer to unraveling the mysteries of high-energy astrophysics and gaining deeper insights into the workings of our universe.

Accretion Disks and High-Energy Emission from Black Holes

In the vast expanse of space, black holes stand as enigmatic objects that continue to captivate the imagination of scientists and enthusiasts alike. These celestial entities, born from the remnants of massive stars, possess a gravitational force so intense that not even light can escape their grasp. As we delve into the mysteries of high-energy astrophysics, one fascinating aspect that demands our attention is the interplay between accretion disks and the emission of high-energy radiation from black holes.

Accretion disks, commonly found around black holes, are swirling disks of gas, dust, and other stellar debris that have fallen under the influence of the black hole's gravitational pull. As these materials orbit the black hole, they gradually spiral inward due to friction and gravitational forces. This process releases an immense amount of energy, giving rise to a wide range of emissions across the electromagnetic spectrum.

At the heart of this phenomenon lies the formation of an accretion disk. When matter falls towards a black hole, it gains angular momentum and forms a disk-like structure around the black hole's event horizon. The inner regions of the disk become incredibly hot and luminous, emitting high-energy radiation such as X-rays and gamma rays.

Understanding the mechanics of accretion disks is crucial in deciphering the nature of high-energy emission from black holes. By studying the spectral properties and variability of these emissions, astrophysicists gain valuable insights into the physical processes occurring within the accretion disk. These observations provide clues about the presence of intense magnetic fields, the role of relativistic particles, and the mechanisms responsible for the production of high-energy photons.

Moreover, accretion disks play a significant role in powering some of the most energetic phenomena in the universe, such as quasars and gamma-ray bursts. Quasars, characterized by their intense luminosity, are believed to be powered by supermassive black holes accreting matter at an astonishing rate. Similarly, gamma-ray bursts, the most energetic explosions in the cosmos, are thought to originate from the collapse of massive stars or the merger of compact objects, leading to the formation of accretion disks around black holes.

The study of accretion disks and high-energy emission from black holes not only contributes to our understanding of the universe but also presents exciting prospects for technological advancements. By harnessing the immense energy released from these cosmic powerhouses, scientists envision the possibility of tapping into new sources of clean and virtually limitless energy.

In conclusion, the exploration of accretion disks and high-energy emission from black holes unravels the intricate secrets of the universe. From their role in powering quasars and gamma-ray bursts to their potential as a source of future energy, these phenomena hold immense significance for astrophysics and humanity's quest to comprehend the cosmos.

Chapter 9: High-Energy Astrophysics and Cosmology

The Role of High-Energy Astrophysics in Cosmology

Astrophysics, the branch of science that seeks to understand the fundamental nature of the universe, has witnessed remarkable advancements over the years. From the discovery of distant quasars to the enigmatic phenomena of gamma-ray bursts, high-energy astrophysics has played a pivotal role in unraveling the mysteries of the cosmos. In this subchapter, we will delve into the significance of high-energy astrophysics in the broader field of cosmology, exploring how it deepens our understanding of the universe we inhabit.

High-energy astrophysics primarily focuses on investigating the most extreme and energetic phenomena occurring in space. By studying sources of high-energy radiation, such as quasars, pulsars, and gamma-ray bursts, astrophysicists gain insights into the physical processes that drive these phenomena. This knowledge not only expands our understanding of the universe but also has profound implications for cosmology, the study of the origin and evolution of the cosmos.

One of the key contributions of high-energy astrophysics to cosmology is the exploration of the early universe. By observing distant quasars, which emit intense radiation, scientists can probe the conditions that prevailed billions of years ago. These observations provide crucial information about the evolution of galaxies and the formation of large-scale structures in the universe.

Additionally, high-energy astrophysics plays a crucial role in deciphering the nature of dark matter and dark energy, two enigmatic entities that dominate the composition and dynamics of the universe. Through the study of cosmic rays and gamma-ray emissions, astrophysicists can indirectly probe the distribution and properties of dark matter. Similarly, the study of supernovae and their high-energy emissions helps us understand the mysterious force driving the accelerated expansion of the universe, known as dark energy.

Furthermore, high-energy astrophysics contributes to our understanding of the fundamental laws of physics. By studying extreme cosmic environments, such as black holes and neutron stars, scientists can test theories of gravity, quantum mechanics, and particle physics under conditions that cannot be replicated on Earth.

In conclusion, high-energy astrophysics plays a crucial role in the field of cosmology, enabling us to unravel the mysteries of the universe. By studying the most energetic phenomena in space, astrophysicists deepen our knowledge of the early universe, dark matter, dark energy, and fundamental physics. This subchapter serves as a gateway to the awe-inspiring world of high-energy astrophysics and its profound implications for our understanding of the cosmos. Whether you are an aspiring astrophysicist or simply curious about the secrets of the universe, this exploration will undoubtedly leave you captivated by the wonders of high-energy astrophysics.

Constraints on Cosmological Models from High-Energy Observations

In the vast expanse of the cosmos, scientists have been tirelessly unraveling the mysteries of high-energy astrophysics. From the enigmatic quasars emitting immense amounts of energy to the explosive gamma-ray bursts that leave astronomers awestruck, these phenomena hold the key to understanding the fundamental workings of our universe. One of the most captivating aspects of this field is how it has helped us place constraints on cosmological models, shedding light on the nature of our existence.

Cosmological models are theoretical frameworks that attempt to explain the origin, evolution, and structure of the universe. They provide a roadmap for scientists to navigate through the vastness of space and time, seeking answers to questions that have baffled humankind for centuries. High-energy observations have played a crucial role in refining and constraining these models, bringing us closer to understanding the intricacies of the cosmos.

One of the remarkable ways in which high-energy observations have shaped cosmological models is through the study of quasars. These incredibly luminous objects, thought to be powered by supermassive black holes at the centers of distant galaxies, emit enormous amounts of energy across the electromagnetic spectrum. By analyzing the properties of quasars, such as their redshift and variability, scientists have been able to gather valuable insights into the expansion rate of the universe, the distribution of matter, and the nature of dark energy.

Another avenue that has provided constraints on cosmological models is the study of gamma-ray bursts (GRBs). These intense explosions, believed to be associated with the deaths of massive stars or the collisions of neutron stars, release an incredible amount of energy in the form of gamma rays. By studying the properties of GRBs, such as their duration and spectrum, scientists have been able to probe the early universe, estimate the rate of star formation, and test theories of gravity.

The combination of high-energy observations from various astrophysical phenomena has allowed researchers to piece together a more complete picture of the cosmos. These observations have challenged existing models and led to the development of new ones, pushing the boundaries of our understanding. As we continue to explore the frontiers of high-energy astrophysics, we can expect more exciting discoveries that will further constrain cosmological models and bring us closer to unraveling the mysteries of our universe.

Whether you are an astrophysics enthusiast or simply curious about the cosmos, understanding the constraints on cosmological models from high-energy observations opens up a fascinating world of knowledge. It invites us to contemplate our place in the universe and marvel at the intricate dance of matter and energy that has shaped everything we see. So join us on this incredible journey as we delve into the depths of high-energy astrophysics and unlock the secrets of the cosmos.

The Evolution of High-Energy Sources in the Universe

Astrophysics is a field of science that has captivated the curiosity of humanity for centuries. From gazing at the stars in wonder to unraveling the mysteries of the cosmos, astrophysicists have strived to understand the intricate workings of the universe. One of the most fascinating aspects of this field is the study of high-energy sources, such as quasars and gamma-ray bursts, which emit vast amounts of energy that defy our imagination.

In this subchapter, we will delve into the evolution of these high-energy sources, tracing their origins and exploring the mechanisms that drive their immense power. Our journey begins with quasars, which are among the most luminous objects in the universe. These enigmatic sources are thought to reside at the centers of distant galaxies and are powered by supermassive black holes. As matter falls into these black holes, it releases an enormous amount of energy in the form of radiation, creating the intense glow of a quasar.

But how do these supermassive black holes form? To answer this question, we must look back in time to the early universe. It is believed that black holes grow through a process called accretion, where surrounding matter is pulled in by the immense gravitational force. As the black hole continues to devour matter, it grows in size, eventually becoming a behemoth capable of powering a quasar.

Moving forward in time, we encounter another class of high-energy sources known as gamma-ray bursts. These extraordinary events release an astonishing amount of energy in the form of gamma rays, the most energetic form of electromagnetic radiation. Gamma-ray bursts are thought to occur during the collapse of massive stars or the collision of neutron stars, unleashing a cataclysmic explosion that briefly outshines entire galaxies.

Studying these high-energy sources provides us with invaluable insights into the processes that shape the universe. By observing their evolution and understanding the mechanisms behind their immense power, we can uncover the secrets of the cosmos and gain a deeper understanding of the fundamental laws of physics.

From quasars to gamma-ray bursts, the study of high-energy astrophysics continues to push the boundaries of human knowledge. It is a field that inspires awe and wonder, captivating the imagination of scientists and enthusiasts alike. As we explore the evolution of these cosmic powerhouses, we inch closer to unraveling the mysteries of the universe and gaining a deeper appreciation for the vastness and complexity of our existence.

Future Prospects and Challenges in High-Energy Astrophysics

As we delve deeper into the mysteries of the universe, the field of high-energy astrophysics continues to expand and captivate the attention of scientists, researchers, and enthusiasts alike. With each new advancement, we gain a better understanding of the extreme phenomena that occur in the cosmos, from quasars to gamma-ray bursts. In this subchapter, we will explore the future prospects and challenges that lie ahead in the realm of high-energy astrophysics.

One of the most exciting prospects in high-energy astrophysics is the advent of new observational instruments and techniques. The development of cutting-edge telescopes, such as the Cherenkov Telescope Array and the James Webb Space Telescope, promises to revolutionize our ability to study high-energy phenomena. These instruments will allow us to probe deeper into the universe, unveiling its most enigmatic secrets. By observing gamma-ray bursts, supermassive black holes, and other high-energy events, we hope to gain insights into the nature of dark matter, the evolution of galaxies, and the origin of cosmic rays.

However, along with these prospects come significant challenges. High-energy astrophysics often deals with phenomena that are rare, transient, and fleeting. Capturing these events requires a combination of luck, skill, and technological prowess. Ensuring that our instruments are sensitive enough to detect these elusive signals poses a formidable task. Moreover, the sheer volume of data generated by these instruments presents a challenge in data analysis and interpretation. Developing sophisticated algorithms and machine learning techniques will be crucial in extracting meaningful information from the vast amount of data collected.

Collaboration and interdisciplinary research will also play a vital role in the future of high-energy astrophysics. The complexities of the field call for expertise from various disciplines, including astrophysics, particle physics, and data science. By fostering collaboration between these disciplines, we can tackle the multifaceted challenges of high-energy astrophysics and make significant advancements in our understanding of the universe.

In conclusion, the future of high-energy astrophysics holds immense promise and excitement. With new observational instruments, advancements in data analysis techniques, and collaborative efforts, we are poised to unravel the mysteries of the cosmos in ways never before imagined. From unraveling the secrets of black holes to deciphering the origins of cosmic rays, high-energy astrophysics continues to push the boundaries of human knowledge. So, whether you are an astrophysics enthusiast or simply curious about the universe, the future prospects and challenges in high-energy astrophysics are sure to captivate your imagination and inspire further exploration.

Chapter 10: The Future of High-Energy Astrophysics

Advancements in Instrumentation and Observation Techniques

In the vast realm of astrophysics, the quest to understand the mysteries of the universe is an ongoing endeavor. It is through constant advancements in instrumentation and observation techniques that we have been able to unravel the secrets of high-energy astrophysics. This subchapter explores the revolutionary strides made in these fields, offering a glimpse into the cutting-edge tools and technologies that have propelled our understanding of the cosmos.

One of the most significant advancements in instrumentation is the development of large telescopes, both on the ground and in space. These behemoths, equipped with state-of-the-art detectors, have expanded our vision beyond what was previously imaginable. Ground-based telescopes, such as the Very Large Telescope (VLT) and the Atacama Large Millimeter/submillimeter Array (ALMA), have provided us with unprecedented resolution and sensitivity. They enable us to observe distant quasars and gamma-ray bursts with remarkable clarity, shedding light on their origins and behavior.

Space-based observatories, on the other hand, offer unique advantages due to their unobstructed view of the cosmos. The Hubble Space Telescope, for instance, has revolutionized our understanding of the universe, capturing breathtaking images of distant galaxies and nebulae. The recent launch of the James Webb Space Telescope promises to push these boundaries even further, with its enhanced capabilities to study the most distant objects in the universe.

Moreover, advancements in detector technology have played a pivotal role in high-energy astrophysics. Emerging technologies, such as the development of solid-state detectors and superconducting sensors, have greatly improved our ability to detect and measure high-energy particles and radiation. These detectors, coupled with ingenious data analysis techniques, allow scientists to decipher the intricate processes occurring in extreme astrophysical environments.

In addition to instrumentation, advancements in observation techniques have revolutionized the field. The concept of multi-wavelength astronomy has been instrumental in unraveling the mysteries of high-energy astrophysics. By studying objects across the electromagnetic spectrum, from radio waves to gamma rays, scientists can piece together a comprehensive picture of the universe. This holistic approach has led to groundbreaking discoveries, such as the identification of black holes and the study of cosmic jets.

In conclusion, advancements in instrumentation and observation techniques have played a pivotal role in our quest to understand the mysteries of high-energy astrophysics. The development of large telescopes, both on the ground and in space, along with cutting-edge detectors and multi-wavelength observations, have allowed us to unravel the secrets of the cosmos. As technology continues to advance, we can only speculate about the exciting discoveries that lie ahead and the profound impact they will have on our understanding of the universe.

Multi-Wavelength and Multi-Messenger Astronomy

In the vast and mysterious universe, the study of high-energy astrophysics has revolutionized our understanding of celestial phenomena. From the distant quasars to the enigmatic gamma-ray bursts, scientists have embarked on a journey to unravel the secrets hidden within the cosmos. One of the most powerful tools at their disposal is the field of multi-wavelength and multi-messenger astronomy.

Traditionally, astronomers have relied on visible light to observe and study celestial objects. However, the universe is not limited to emitting only visible light. It is a rich tapestry of various wavelengths, each carrying unique information about the nature and behavior of these cosmic wonders. Multi-wavelength astronomy expands our view beyond the visible spectrum, allowing us to explore the universe in infrared, ultraviolet, X-ray, and even gamma-ray wavelengths.

By observing the cosmos at different wavelengths, scientists can discern the various processes and phenomena occurring within astrophysical systems. For example, X-ray observations can reveal the presence of high-energy particles or the presence of black holes, while infrared observations can unveil the formation of new stars in dusty regions. Combining these different datasets provides a more comprehensive picture of the universe, enabling us to connect the dots and uncover the intricacies of astrophysical processes.

But the universe doesn't only speak in terms of electromagnetic radiation; it also communicates through cosmic messengers. Enter multi-messenger astronomy, which utilizes different types of particles, such as neutrinos and gravitational waves, to probe the mysteries of the cosmos. Neutrinos, elusive subatomic particles, can traverse vast distances without being affected by magnetic fields or other matter, making them powerful cosmic messengers. Gravitational waves, on the other hand, are ripples in space-time caused by the most energetic events in the universe, such as the merging of black holes or neutron stars.

By combining the information from these cosmic messengers with the observations obtained from multi-wavelength astronomy, scientists can gain a deeper understanding of the most extreme and violent events in the universe. These include the birth and death of stars, the dynamics of black holes, and the behavior of exotic phenomena like gamma-ray bursts.

Multi-wavelength and multi-messenger astronomy represent the cutting edge of astrophysics, allowing us to explore the universe in ways unimaginable just a few decades ago. As our instruments and technologies continue to advance, we can eagerly anticipate the discovery of new cosmic marvels and the unraveling of the mysteries that lie within our awe-inspiring universe. Whether you are an astrophysics enthusiast or a curious individual seeking to understand the wonders of our cosmos, the field of multi-wavelength and multi-messenger astronomy offers an exciting and ever-expanding frontier of knowledge waiting to be explored.

Theoretical Advances and Computational Modeling

In the vast and ever-evolving field of high-energy astrophysics, theoretical advances and computational modeling play a crucial role in unraveling the mysteries of the universe. This subchapter delves into the fascinating world of theoretical astrophysics and the powerful tools of computational modeling that enable scientists to explore the enigmatic phenomena such as quasars and gamma-ray bursts.

Theoretical advances in astrophysics involve the development and refinement of mathematical models and theories that explain the behavior and origin of celestial objects and phenomena. These theoretical frameworks provide a foundation for understanding complex astrophysical processes, allowing scientists to make predictions, test hypotheses, and interpret observational data. From the inception of general relativity to the formulation of quantum field theories, theoretical advances have revolutionized our understanding of the cosmos.

Computational modeling, on the other hand, allows scientists to simulate and replicate astrophysical phenomena using powerful computers. By solving complex mathematical equations numerically, researchers can simulate the behavior of celestial objects, study the dynamics of galaxies, and even recreate the conditions of the early universe. Computational models provide a valuable tool for validating theoretical predictions, interpreting observational data, and exploring scenarios that are difficult or impossible to recreate in a laboratory.

This subchapter explores various aspects of theoretical advances and computational modeling in high-energy astrophysics. It discusses the theoretical frameworks that have shaped our understanding of quasars, enigmatic cosmic objects that emit enormous amounts of energy. We delve into the models explaining the formation and evolution of these celestial powerhouses, as well as the mechanisms that drive their incredible luminosity.

Furthermore, we explore the theoretical advances and computational models that shed light on gamma-ray bursts, the most energetic explosions in the universe. We delve into the different proposed mechanisms for these bursts, ranging from the collapse of massive stars to the collision of compact objects. Through theoretical advancements and computational modeling, scientists have made significant progress in unraveling the origins and physics behind these awe-inspiring events.

By delving into theoretical advances and computational modeling, this subchapter aims to provide a glimpse into the cutting-edge research and methodologies employed in the field of high-energy astrophysics. Although the content may be technical, it aims to make these fascinating concepts accessible to everyone interested in the mysteries of the universe. Whether you are an astrophysics enthusiast or a curious reader, this subchapter will broaden your understanding of the theoretical foundations and computational tools that pave the way for unraveling the enigmas of high-energy astrophysics.

Collaborative Efforts and International Initiatives

In the field of astrophysics, the pursuit of knowledge knows no boundaries or borders. Scientists and researchers from around the world come together in collaborative efforts and international initiatives to unravel the mysteries of high-energy astrophysics. These collective endeavors have played a significant role in advancing our understanding of the universe and its most enigmatic phenomena.

Collaboration lies at the heart of scientific progress. The challenges posed by high-energy astrophysics are complex and multifaceted, requiring the expertise and resources of multiple institutions and countries. By pooling their knowledge and capabilities, researchers can tackle these challenges more effectively and achieve breakthrough discoveries.

One of the most notable international initiatives in the field is the Global Relay of Observatories Watching Transients Happen (GROWTH) project. GROWTH brings together scientists and observatories from all corners of the globe to observe and study transient astronomical events, such as supernovae and gamma-ray bursts. By coordinating observations across different time zones and continents, GROWTH maximizes the amount of data collected and provides a more comprehensive understanding of these dynamic phenomena.

International collaboration also extends to the development of cutting-edge technology and instruments. Large-scale projects like the Square Kilometre Array (SKA) bring together multiple countries to build the world's most powerful radio telescope. By combining the expertise and resources of participating nations, the SKA will revolutionize our understanding of the cosmos, enabling us to probe deeper into the universe than ever before.

These collaborative efforts not only advance scientific knowledge but also foster a sense of global community among astrophysicists. Researchers from diverse backgrounds and cultures come together, exchange ideas, and learn from one another. This cultural exchange enhances creativity and innovation, leading to novel approaches and perspectives in high-energy astrophysics.

Furthermore, international initiatives in astrophysics have a broader impact beyond the scientific community. They inspire the next generation of scientists and promote scientific literacy among the general public. By showcasing the power of collaboration and the wonders of the universe, these efforts ignite curiosity and encourage young minds to pursue careers in astrophysics.

In conclusion, collaborative efforts and international initiatives are essential in unraveling the mysteries of high-energy astrophysics. These endeavors bring together scientists, observatories, and nations to tackle complex challenges, develop cutting-edge technology, and foster a global community of astrophysicists. By working together, we can push the boundaries of knowledge and unlock the secrets of the universe, ultimately benefiting everyone and advancing our understanding of the cosmos.

Conclusion: Unveiling the Mysteries of High-Energy Astrophysics

Summary of Key Findings

In the book "From Quasars to Gamma-Ray Bursts: Unraveling the Mysteries of High-Energy Astrophysics," the author delves into the fascinating field of astrophysics, exploring the enigmatic phenomena that occur within our vast universe. This subchapter, "Summary of Key Findings," serves as a comprehensive overview of the book's most significant discoveries and breakthroughs, aimed at both general readers and those with a particular interest in astrophysics.

One of the primary areas of focus in this book is the study of quasars, which are incredibly bright and distant celestial objects. Through extensive research and observation, scientists have discovered that quasars are powered by supermassive black holes at the centers of galaxies. These black holes devour surrounding matter, emitting enormous amounts of energy in the process. Understanding the mechanisms behind quasars has provided valuable insights into the formation and evolution of galaxies.

Another major topic explored in this book is gamma-ray bursts (GRBs), which are intense explosions of high-energy radiation. Scientists have made significant progress in unraveling the mysteries surrounding these powerful events. They have identified two types of GRBs: short-duration bursts resulting from the merger of neutron stars, and long-duration bursts associated with the collapse of massive stars. The detection and analysis of GRBs have shed light on the nature of the early universe, the formation of black holes, and the production of heavy elements.

Furthermore, the book delves into the study of cosmic rays, which are energetic particles that originate from various astrophysical sources. Scientists have successfully identified cosmic rays with energies exceeding any produced on Earth, leading to the discovery of extremely energetic events such as supernovae and active galactic nuclei. The understanding and characterization of cosmic rays provide crucial insights into the underlying physical processes that occur within these astrophysical sources.

Additionally, the author explores the field of high-energy astrophysics using cutting-edge observational techniques and sophisticated instruments. The book emphasizes the importance of space-based observatories, such as the Hubble Space Telescope and the Chandra X-ray Observatory, in capturing high-resolution images and collecting invaluable data from distant objects.

Overall, "From Quasars to Gamma-Ray Bursts: Unraveling the Mysteries of High-Energy Astrophysics" provides a comprehensive summary of the key findings in the field of astrophysics. Whether you are a general reader seeking to expand your knowledge of the universe or an astrophysics enthusiast looking for the latest discoveries, this subchapter offers a concise yet informative overview of the fascinating world of high-energy astrophysics.

The Impact of High-Energy Astrophysics on our Understanding of the Universe

Astrophysics, the branch of science that deals with the study of celestial objects and the phenomena occurring in space, has witnessed remarkable advancements in recent decades. In particular, the field of high-energy astrophysics has played a pivotal role in unraveling the mysteries of the universe. From quasars to gamma-ray bursts, the exploration of high-energy phenomena has revolutionized our understanding of the cosmos, captivating astrophysics enthusiasts and professionals alike.

One of the most significant impacts of high-energy astrophysics is the discovery and exploration of quasars. These enigmatic objects, characterized by their extraordinary luminosity and massive black holes at their cores, have shed light on the early universe and the processes that govern galaxy formation. By studying the high-energy emissions emitted by quasars, scientists have been able to trace the evolution of galaxies and uncover the intricate web of cosmic structures that make up our universe.

Another intriguing aspect of high-energy astrophysics is the study of gamma-ray bursts (GRBs). These intense bursts of gamma-ray radiation, lasting from a few milliseconds to several minutes, have posed a perplexing puzzle for astrophysicists. Through dedicated observations and analysis, scientists have made significant progress in unraveling the origins of these powerful explosions. GRBs are now believed to be associated with the collapse of massive stars or the merger of compact objects such as neutron stars or black holes. The study of GRBs has not only deepened our understanding of stellar evolution but has also provided valuable insights into the fundamental physics of extreme environments.

Moreover, high-energy astrophysics has enabled scientists to explore the mysterious phenomenon of dark matter. Although invisible, dark matter is thought to constitute a significant portion of the universe's mass. By studying the high-energy emissions associated with the annihilation or decay of dark matter particles, researchers have made significant strides in constraining the properties and distribution of this elusive substance. These findings have not only refined our understanding of the universe's composition but have also paved the way for future discoveries in particle physics and cosmology.

In conclusion, the field of high-energy astrophysics has had a profound impact on our understanding of the universe. Through the study of quasars, gamma-ray bursts, and dark matter, scientists have made remarkable strides in unraveling cosmic mysteries and pushing the boundaries of human knowledge. The captivating nature of high-energy astrophysics continues to inspire and engage both astrophysics enthusiasts and professionals, fostering a deeper appreciation for the wonders of the cosmos.

Final Thoughts and Future Directions

As we come to the end of this captivating journey through the mysteries of high-energy astrophysics, it is important to reflect on the knowledge we have gained and consider the exciting possibilities that lie ahead. From quasars to gamma-ray bursts, our exploration of the universe's most energetic phenomena has provided us with a deeper understanding of the cosmos and opened up new avenues for scientific exploration.

Throughout this book, we have delved into the fascinating world of astrophysics, uncovering the secrets of quasars, the most luminous objects in the universe. We have learned how these enigmatic cosmic beacons are powered by supermassive black holes, devouring surrounding matter and releasing enormous amounts of energy in the process. Our understanding of quasars has not only shed light on the early universe but has also challenged our existing knowledge of black holes and their role in galaxy evolution.

The journey has also taken us to the thrilling realm of gamma-ray bursts, the most energetic explosions known to occur in space. These intense bursts of gamma-ray radiation have puzzled scientists for decades, and their origins still remain partially elusive. However, recent advancements in observational techniques and theoretical models are bringing us closer to unraveling the mysteries surrounding these cosmic fireworks. By studying the afterglows of gamma-ray bursts and their host galaxies, we hope to gain valuable insights into the processes that drive these cataclysmic events and their profound impact on the surrounding cosmic environment.

Looking to the future, the field of high-energy astrophysics holds immense promise. Advances in technology, such as the development of more sensitive telescopes and detectors, will enable us to explore the universe in even greater detail. We can anticipate the discovery of new astrophysical phenomena and the refinement of existing theories as we continue to push the boundaries of our understanding.

Moreover, collaborations between astrophysicists and other scientific disciplines will foster interdisciplinary approaches to tackle complex astrophysical problems. The synergy between astrophysics and fields such as particle physics and cosmology will provide us with a comprehensive understanding of the universe on both the largest and smallest scales.

In conclusion, our journey through the mysteries of high-energy astrophysics has been awe-inspiring, revealing the astonishing power and beauty of the cosmos. As we move forward, driven by curiosity and guided by science, we can look forward to a future where the secrets of quasars, gamma-ray bursts, and other high-energy phenomena are unraveled. The universe continues to beckon us, inviting us to explore its deepest recesses and expand the frontiers of human knowledge. Let us embark on this exciting adventure together, for the wonders of the cosmos are waiting to be discovered by everyone.